Stark County District Library
www.StarkLibrary.org
330.452.0665

S0-EIK-918

Full STEAM Ahead!

Math Matters

Telling Time Together

Adrianna Morganelli

Title-Specific Learning Objectives:

Readers will:

- Identify digital and analog clocks and describe how they measure time.
- Tell time using digital and analog clocks.
- Use the pictures in the book to describe how to tell time.

High-frequency words (grade one)	Academic vocabulary
after, and, at, is, it, the, to, what, where	a.m., analog, colon, digital, hands, hour, minute, p.m.

Before, During, and After Reading Prompts:

Activate Prior Knowledge and Make Predictions:
Show children the cover images and read the title out loud. Ask children:

- How do people tell time? What tools do they use?
- Do you know what time it is now? How do you know?

Allow children to share their thoughts. Invite them to draw pictures of clocks in small groups. Have children share their pictures and explain what they drew.

During Reading:
After reading pages 6 and 7, ask children to review the pictures and labels on the digital and analog clocks.

Ask them:

- What are the parts of an analog clock? How do they measure time?
- What are the parts of a digital clock? How do they measure time?

After Reading:
Review the pictures children drew before reading the book. Do any of the pictures need to be changed to match the clocks in the book? Make corrections. Then, play a bingo game with children. Create bingo sheets with both digital and analog clocks on them. Pass the cards out to students and call out times, such as "ten thirty." When a child has made a line, review their answers with the whole group.

Author: Adrianna Morganelli

Series Development: Reagan Miller

Editor: Janine Deschenes

Proofreader: Melissa Boyce

STEAM Notes for Educators: Janine Deschenes

Guided Reading Leveling: Publishing Solutions Group

Cover, Interior Design, and Prepress: Samara Parent

Photo research: Janine Deschenes and Samara Parent

Production coordinator: Katherine Berti

Photographs:
iStock: SDI Productions: p. 20 (r)
Shutterstock: S-F: p. 12 (r); Anton_Ivanov: p. 16 (r)

All other photographs by Shutterstock

Library and Archives Canada Cataloguing in Publication

Title: Telling time together / Adrianna Morganelli.
Names: Morganelli, Adrianna, 1979- author.
Description: Series statement: Full STEAM ahead!
Identifiers: Canadiana (print) 20190231718 |
Canadiana (ebook) 20190231734 |
ISBN 9780778772705 (softcover) |
ISBN 9780778772217 (hardcover) |
ISBN 9781427124630 (HTML)
Subjects: LCSH: Time—Juvenile literature. |
LCSH: Clocks and watches—Juvenile literature.
Classification: LCC QB209.5 .M67 2020 | DDC j529—dc23

Library of Congress Cataloging-in-Publication Data

Names: Morganelli, Adrianna, 1979- author.
Title: Telling time together / Adrianna Morganelli.
Description: New York, New York : Crabtree Publishing Company, [2020] | Series: Full steam ahead! | Includes index.
Identifiers: LCCN 2019052918 (print) | LCCN 2019052919 (ebook) | ISBN 9780778772217 (hardcover) | ISBN 9780778772705 (paperback) | ISBN 9781427124630 (ebook)
Subjects: LCSH: Time--Juvenile literature. | Clocks and watches--Juvenile literature.
Classification: LCC QB209.5 .M66 2020 (print) | LCC QB209.5 (ebook) | DDC 529--dc23
LC record available at https://lccn.loc.gov/2019052918
LC ebook record available at https://lccn.loc.gov/2019052919

Printed in the U.S.A./032020/CG20200127

Table of Contents

Crabtree Publishing Company
www.crabtreebooks.com   1-800-387-7650
 In Canada: We acknowledge the financial support of the Government of Canada through the Canada Book Fund for our publishing activities.

Published in Canada
Crabtree Publishing
616 Welland Ave.
St. Catharines, Ontario
L2M 5V6

Published in the United States
Crabtree Publishing
PMB 59051
350 Fifth Avenue, 59th Floor
New York, New York 10118

Published in the United Kingdom
Crabtree Publishing
Maritime House
Basin Road North, Hove
BN41 1WR

Published in Australia
Crabtree Publishing
Unit 3 – 5 Currumbin Court
Capalaba
QLD 4157

An Exciting Day

Charlotte is excited! Today her class is going to the science center. Charlotte's alarm clock just woke her up.

Charlotte's alarm clock wakes her up at seven o'clock in the morning.

We use clocks to tell time. There are two types of clocks. They are **digital** clocks and **analog** clocks.

A digital clock shows numbers on a screen. Charlotte's alarm clock is a digital clock.

This is an analog clock. It has moving hands that point to numbers.

Hours and Minutes

Clocks show time in hours and minutes.
One minute is 60 seconds long.
One hour is 60 minutes long.

The hands on analog clocks measure hours, minutes, and seconds.

Some clocks have a hand that measures seconds. It takes 60 seconds, or one minute, for the second hand to go all the way around the clock.

The hour hand points to the hour. It is shorter than the minute hand.

The minute hand points to the minute. Each dot on the clock is one minute.

When the minute hand has gone all the way around the clock, 60 minutes, or one hour, has passed.

Charlotte woke up at seven o'clock. She washed her hair, brushed her teeth, and got dressed. Now, it is seven thirty. How many minutes have passed?

30 minutes have passed. 30 minutes is also half of one hour.

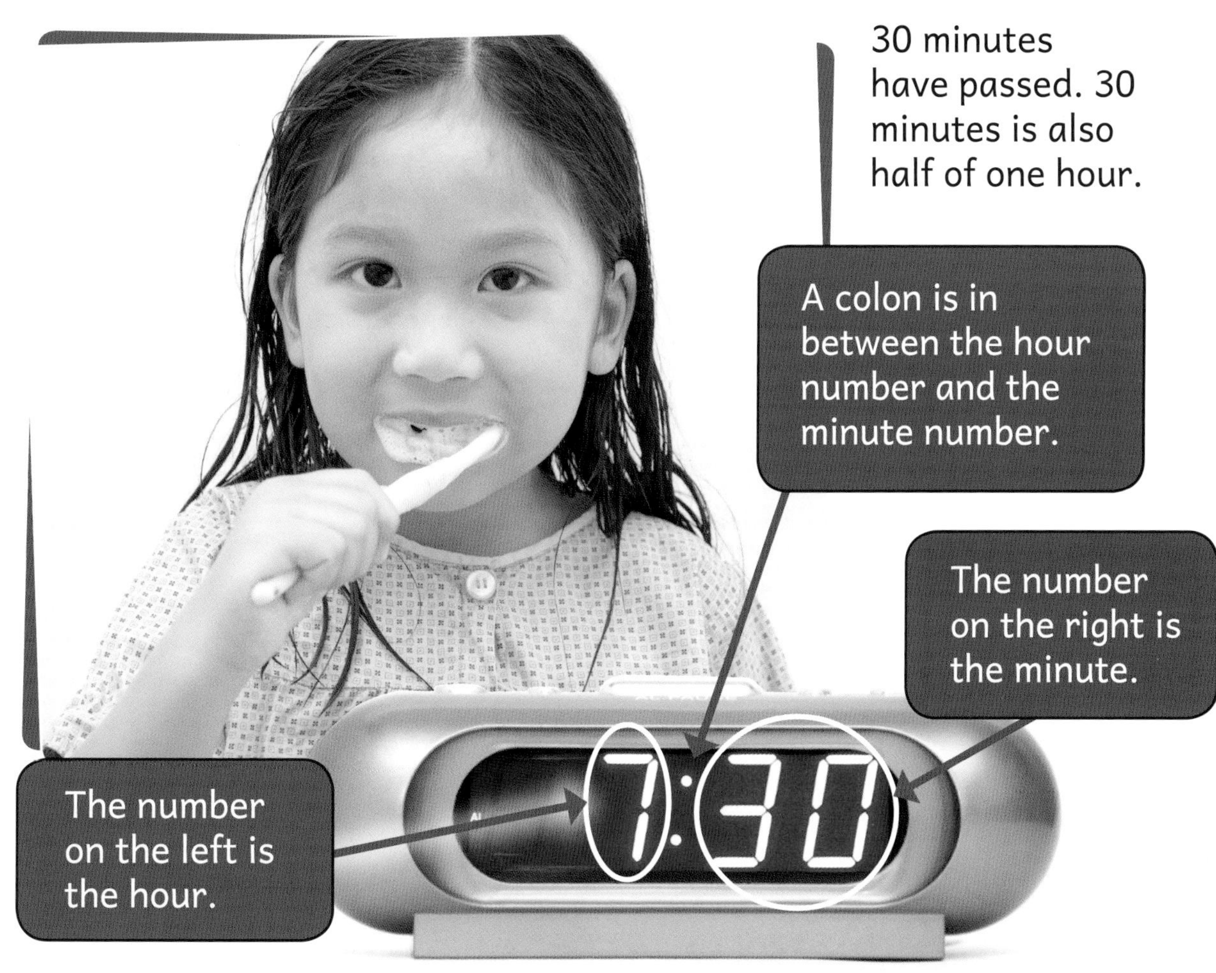

Digital clocks use numbers to show hours and minutes.

Time for School

Charlotte puts on her favorite shirt and eats breakfast. “School starts at eight thirty,” says Charlotte’s mom. “You should finish eating.” Charlotte looks at the clock. Half an hour has passed. What time is it?

Is the clock in the kitchen a digital clock or an analog clock?

The clock shows that it is eight o'clock. Charlotte takes her last bite. She and her mom walk to school together.

Charlotte hugs her mom goodbye. Then she checks her watch. Their walk to school took 30 minutes. What time is it now?

Using an Analog Clock

Charlotte waves to her friend Reno in the classroom. They can't wait to leave for the science center! Their teacher tells them that the bus will be picking them up at nine o'clock. "What time is it now?" asks Reno.

Look at the clock. Where is the hour hand pointing? Where is the minute hand pointing? Do you know the time?

The numbers 1 to 12 are on an analog clock. The space between each number **represents** five minutes. When the minute hand points to 6, 30 minutes have passed. When the minute hand points to 12, one hour has passed.

At eight thirty, the hour hand will be halfway between 8 and 9. The minute hand will point to 6. That is halfway around the clock.

At nine o'clock, the hour hand will point to 9. The minute hand will point to 12. That is all the way around the clock.

Using a Digital Clock

At the science center, there is so much to see. There is a special experiment for the students at eleven o'clock. Charlotte can't wait. She wonders what time it is now.

Look at the watch. Can you help Charlotte tell the time?

Digital clocks show hours with numbers from 1 to 12. They show minutes with numbers from 1 to 59.

Half an hour is 30 minutes. A digital clock shows half an hour with the number 30.

One hour is 60 minutes. After 10:59, the clock changes to 11:00. The hour number **increases** by one. The minute number becomes 00.

The students try the experiment. They have fun working together to mix **materials**. The **mixture** overflows!

Morning to Afternoon

After the experiment, Charlotte is hungry. “Lunch is at noon, or twelve o’clock p.m.,” the teacher says. “What does p.m. mean?” asks Reno. “It means that morning is over. It is afternoon now,” explains the teacher.

The digital clock and the analog clock both show twelve o’clock. Time for lunch!

There are 24 hours in a day. The 12 morning hours are labeled a.m. The 12 hours after noon are labeled p.m.

The a.m. hours start at midnight, in the middle of the night. That is 12:00 a.m. They end at 11:59 a.m.

The p.m. hours start at noon, in the middle of the day. That is 12:00 p.m. They end at 11:59 p.m.

What time are these children playing outside? The clock says three o'clock. Three o'clock a.m. is three hours after midnight. It is dark outside at that time. These children are playing outside at three o'clock p.m.

Counting Minutes

After lunch, Charlotte and Reno are excited to explore the space **exhibit**. "Look!" says Reno. "A sign says that an astronaut will give a **presentation** at one o'clock!"

What time is it when Reno notices the sign?

"How many minutes until it is time for the presentation?" asks Charlotte.

It is twelve thirty. The hour hand is between 12 and 1. The minute hand points to 6.

At one o'clock, the hour hand will point to 1. The minute hand will have moved from 6 to 12. That is halfway around the clock, or 30 minutes.

Reno can hardly wait 30 minutes for the presentation! He will get a turn on this spinning machine. It helps astronauts learn how to balance in space.

How Long Did It Take?

After the presentation, Charlotte wants to go to the activity room. The friends have fun building cool objects with blocks. "I'm building an alien from space!" says Charlotte.

It is now two o'clock. How long was the presentation?

Charlotte and Reno build many objects. “Wow!” says Reno. “How long did that take us?” The children look at the wall clock.

Look at the clock. Where is the hour hand pointing? Where is the minute hand pointing? How long have Charlotte and Reno been building?

The End of the Day

The teacher looks into the activity room. "The bus will pick us up in half an hour," he reminds the students. "How much time do we have left?" asks Reno.

Look at the clock. What time will it be in half an hour? How many minutes will pass?

At bedtime that night, Charlotte tells her mom all about the trip. “My favorite part was the cool experiment. The mixture went everywhere!” she exclaims. “That’s exciting!” her mom replies. “But it’s time to sleep.”

What time is Charlotte’s bedtime? Is the time a.m. or p.m.? What time is your bedtime?

Words to Know

analog [AN-a-log] adjective Something that shows information with constantly changing positions, such as the hands on a clock

digital [DIJ-i-tl] adjective Something that shows digits, or numbers, such as a digital clock

exhibit [ig-ZIB-it] noun A collection of objects on display

increases [in-KREES-es] verb Goes up or makes greater

materials [m*uh*-TEER-ee-*uh*ls] noun Things from which something is made

mixture [MIKS-cher] noun Two or more materials mixed together

presentation [prez-*uh* n-TEY-sh*uh* n] noun A performance or talk in which something is shown to an audience

represents [rep-ri-ZENTs] verb Stands for or takes the place of

A noun is a person, place, or thing.

A verb is an action word that tells you what someone or something does.

An adjective is a word that tells you what something is like.

Index

About the Author

Adrianna Morganelli is an editor and writer who has worked with Crabtree Publishing on countless book titles. She is currently working on a children's novel.

To explore and learn more, enter the code at the Crabtree Plus website below.

www.crabtreeplus.com/fullsteamahead

Your code is:
fsa20

STEAM Notes for Educators

Full STEAM Ahead is a literacy series that helps readers build vocabulary, fluency, and comprehension while learning about big ideas in STEAM subjects. *Telling Time Together* uses pictures of digital and analog clocks to help readers understand how time is measured. The STEAM activity below helps readers extend the ideas in the book to build their skills in math and science.

Changes Over Time

Children will be able to:

- Connect daily activities with the time they happen.
- Make observations and record the appearance of the sky at certain times.
- Correctly record time on digital and analog clocks.

Materials

- Observations Over Time Worksheet
- Observations Over Time Completed Example

Guiding Prompts

After reading *Telling Time Together*, ask:

- Can you point to an analog clock? How does an analog clock measure time?
- Can you point to a digital clock? How does a digital clock measure time?

Activity Prompts

Have a discussion about how scientists make observations. Ask children why observations are important. Then, ask them how telling time relates to making observations. Lead children to understand that scientists must record the time when they observe how something changes over time.

Tell children that they will practice this skill in this activity. They will act like scientists who are learning about how the sky changes. Like scientists, they will record what they see at different times. They need to show what they learned about measuring time on digital and analog clocks.

Hand each child an Observations Over Time Worksheet. Allow them time to review the activity and read the instructions. Then, review the activity together. Show them the Observations Over Time Completed Example.

When children have completed the activity, have them compare their observations with their peers. Discuss what the children observed at each time.

Extensions

- Invite children to make observations over a longer period of time, such as a week, and draw conclusions about how the sky usually looks at each time.

To view and download the worksheets, visit **www.crabtreebooks.com/resources/printables** or **www.crabtreeplus.com/fullsteamahead** and enter the code **fsa20**.